Your First

TARANTULA

CONTENTS

Introduction	4
Basic Biology	8
Buying a Tarantula	10
Popular Tarantula Species	11
Housing and Equipment	22
Feeding	26
Moulting and Sexing	27
Health Matters	30
Final Remarks	32
Bibliography	32
Useful Addresses	33

The author would like to thank Mark Kent for his help with proof-reading and also for supplying some of the spiders who appear in this book.

Photographs by:
Russell Willis

Front cover painting by:
D A Lish

t.f.h.
KINGDOM

©1999 by Kingdom Books PO9 5TL ENGLAND

INTRODUCTION

Many insects and invertebrates have a stigma attached to them but few animals have the ability to evoke such a strong reaction as the spider. Whatever the reasons for this reaction, irrational or otherwise, the fact remains that spiders are not always the most affectionately regarded of animals.

To many people the idea of keeping a spider as a pet is distinctly odd. This is hardly surprising given the image that spiders, and particularly 'tarantulas', have. Due to the proliferation of fictional and grossly exaggerated information in the media, often concerning their toxic capabilities, spiders are treated with contempt and regarded as fearsome lowly creatures by a large proportion of the general public. Yet there are a growing number of people who have discovered just how fascinating these creatures are and, as a result, spiders are now more commonly kept as pets than ever before.

The principle aim of this guide is to serve as an introduction to a very interesting and immensely rewarding hobby; and, secondly, to provide all the information a beginner needs in order to select, house and look after some of the world's largest and most exotic arachnids.

The tarantula is perhaps the most infamous of all spiders, yet the term 'tarantula' itself is somewhat of a misnomer. The true tarantula (*Lycosa tarantula*) is a relatively unspectacular little spider found in the Mediterranean region. The true tarantula acquired its name from a dance originating in the town of Taranto, Italy.

A beautiful sub-adult Mexican Fire Leg (*Brachypelma boehmei*). 'Sub-adult' refers to a spider in the final stage before adulthood.

YOUR FIRST TARANTULA — INTRODUCTION

Avicularias are the epitome of big, hairy spiders. Members of this genus make an excellent first purchase and should be housed in a tall terrarium.

When this dance was performed, it was said to relieve the madness and delirium caused by the spider's bite. The dance, known as the 'tarantella', gave the tarantula its name. Today this term is broadly applied to many of the large tropical spiders although, technically, tarantulas should be referred to as 'theraphosids', belonging to the sub-order Mygalomorphae.

Not only is the term 'tarantula' strictly incorrect, but the other commonly used name - 'bird eater' - is also exaggerated and misused. The term 'bird eater' is derived from the accounts of early naturalists and explorers who encountered giant spiders on their travels. Henry Walter Bates was one such naturalist who, in the 1800s, described birds being caught in webs of South American spiders. While these accounts have captured peoples' interest and helped create a sensational image of terrifying eight-legged beasts, the likelihood of birds forming any part of a spider's staple diet is unlikely.

Information concerning the tarantula's bite has also been somewhat distorted. In fact, if one is unlucky enough to be bitten, the result is likely to be comparable with that of a bee sting, although there are a few exceptions and an allergic reaction to the venom of certain species could be more serious.

INTRODUCTION   YOUR FIRST TARANTULA

Many New World tarantulas will brush fine hairs from their abdomen when threatened. A bald abdominal patch is a sure sign of a hair kicker. (Photo: Mexican Red-Knee.)

Whilst many tarantulas may look threatening, mainly due to their large proportions, they fall a long way short of being the most dangerous spiders in the world. Indeed, the most toxic species are also some of the most innocuous looking. These include the Black Widow (*Latrodectus*); Wandering Spiders (*Ctenidae*); Recluse Spiders (*Loxosceles*), South American Wolf Spiders (*Lycosa*) and Funnel Web Spiders belonging to the genus *Atrax*. To the layperson, all of these species appear quite harmless when compared to the more sinister-looking theraphosids (tarantulas); The size of the spider is not directly proportional to its toxicity.

Many theraphosids have other means of defending themselves and will only bite as a last resort. Warning colouration is an effective deterrent to a would-be predator and is sometimes exhibited in the form of vivid leg bands (this is particularly noticeable in species such as The Mexican Red-Knee (*Brachypelma smithi*). Fine hairs can also be brushed from the abdomen of some New World species which, when caught in the eyes of a potential predator, would undoubtedly cause severe irritation.

In the wild, tarantula and man cross paths fairly infrequently and tarantulas are generally secretive creatures who enjoy the safety of their own burrow or web. However, in their native country, tarantulas are often actively encouraged to dwell alongside humans. A free-range tarantula living in a dark corner or under the tin roof of a hut can be an efficient form of pest control, feeding on cockroaches and other insect pests.

In certain parts of the world spiders actually form part of the diet of indigenous people, not only providing a valuable source of protein but also being relished as a delicacy. The Thai Black (*Haplopelma minax*) - *minax* meaning 'edible' - is one such species, originating from Thailand and Burma.

There are thought to be around 35,000 species of spider in existence, ranging in size from just a few millimetres to spectacular giants of enormous proportions.

While some are cryptically camouflaged as flowers or go unnoticed among vegetation, others blatantly advertise their toxic attributes in the form of vivid, eye-catching warning colours. If one takes a closer look at these creatures it is hard not to appreciate the almost unbelievable variation of colours, shapes and forms.

Occupying habitats as diverse as tropical rain forests, deserts, and even some aquatic environments, this remarkably successful group of creatures deserves more than our fear and loathing. Ultimately, spiders should demand our respect.

(Above and below) The Peru Blonde (*Lasiodorides* sp.) is a large and beautiful species. Although somewhat reluctant to bite, it will readily expose its sizeable fangs and kick hairs when disturbed.

BASIC BIOLOGY

The class Arachnida comprises the spiders, scorpions, palpigrades, harvestmen, mites, ticks, pseudoscorpions, tail-less whip scorpions, whip scorpions and wind scorpions.

Tarantulas belong to the sub-order Mygalomorphae which can be further divided into numerous families. Theraphosidae is the family that one refers to when talking about 'tarantulas' and 'bird eaters', of which there are approximately 650-700 identified species with undoubtedly many more waiting to be discovered.

Spiders are instantly recognisable to most people. Instead of possessing three body segments like insects, spiders have two distinct body sections: the prosoma and the opisthosoma (the cephalothorax and abdomen, as illustrated on page 9).

The cephalothorax is essentially a fused head and thorax to which the legs, mouth-parts and abdomen are attached. The upper surface is known as the carapace and the eyes are situated on a small raised structure towards the front.

The abdomen acts as a food store and houses most of the spider's internal organs such as the reproductive organs, the heart, book lungs and much of the digestive tract.

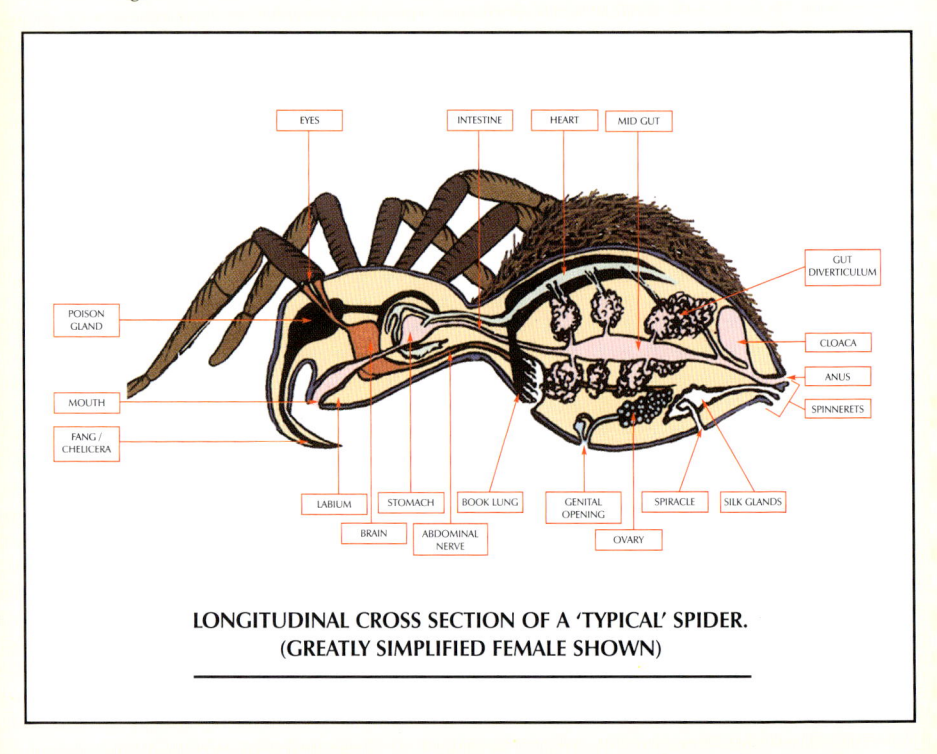

**LONGITUDINAL CROSS SECTION OF A 'TYPICAL' SPIDER.
(GREATLY SIMPLIFIED FEMALE SHOWN)**

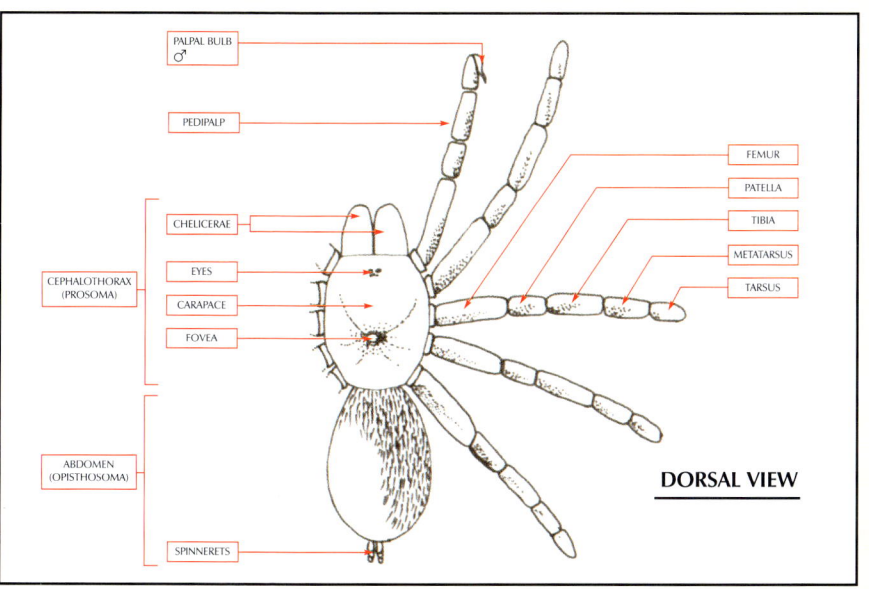

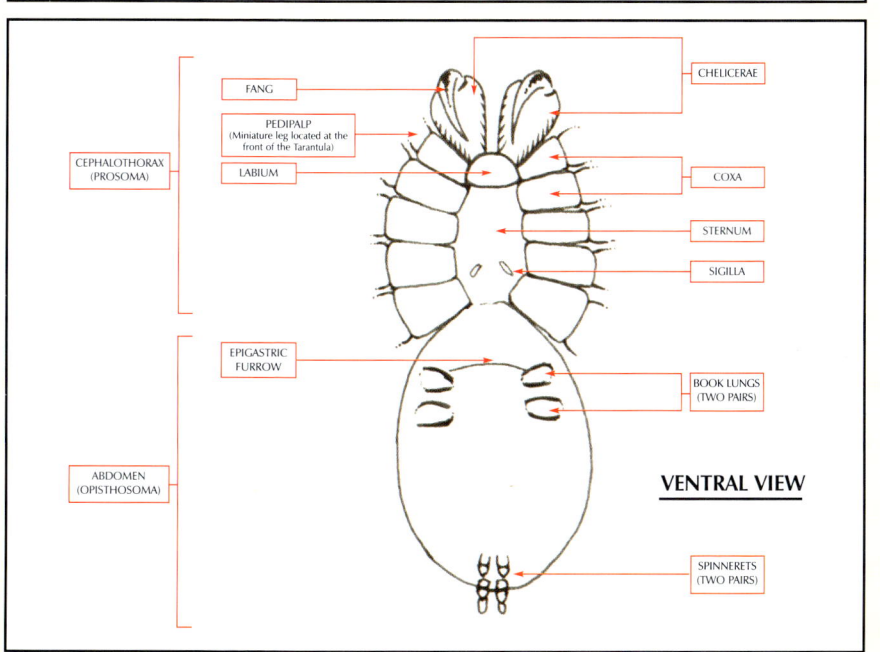

YOUR FIRST TARANTULA   BASIC BIOLOGY

DORSAL VIEW

Labels: PALPAL BULB ♂, PEDIPALP, FEMUR, PATELLA, TIBIA, METATARSUS, TARSUS, CHELICERAE, EYES, CARAPACE, FOVEA, CEPHALOTHORAX (PROSOMA), ABDOMEN (OPISTHOSOMA), SPINNERETS

VENTRAL VIEW

Labels: FANG, CHELICERAE, PEDIPALP (Miniature leg located at the front of the Tarantula), LABIUM, COXA, STERNUM, SIGILLA, CEPHALOTHORAX (PROSOMA), EPIGASTRIC FURROW, BOOK LUNGS (TWO PAIRS), ABDOMEN (OPISTHOSOMA), SPINNERETS (TWO PAIRS)

BUYING A TARANTULA

Before buying a tarantula it is good idea to examine one's motives and identify the role the new pet is to fulfil. Tarantulas cannot be stroked or petted and therefore cannot be considered as being a 'pet' in the conventional sense. Although it is possible to handle certain species, this is to be strongly discouraged. Other than giving the handler a quick thrill, the benefits to the spider are somewhat limited.

Often there is a greater risk of damaging the spider than there is of the spider biting the handler; a tarantula's abdomen is particularly fragile and even a minor fall can cause it to rupture. The golden rule concerning handling is very simple - don't!

As a newcomer to the field, it is critical that a suitable species is selected as not all tarantulas make for a good first purchase. There is great variation in the temperament not only of different species but also between individuals. Many of the African Baboon spiders, for example, tend to be extremely aggressive and will rear up (exposing their fangs) and lunge repeatedly at the slightest of movements. This behaviour sets one's heart racing and is likely to cause an inexperienced keeper to break into a cold sweat! On the other hand, there are species that rarely show aggression, even if severely provoked. These species are the ones that a beginner should consider as a first buy.

Tarantulas are commonly seen for sale in pet shops although the selection on offer may be somewhat limited. A specialist entomological dealer or breeder will generally stock a much wider variety. Livestock, foods and equipment can usually be sent by courier or mail. A good way of obtaining specimens and sharing knowledge is to join a club or society. Members will often have surplus offspring from breeding projects and, consequently, a young, captive-bred spider can be purchased very inexpensively.

Tarantulas are still collected from the wild to satisfy the trade. Provided that only small numbers are imported then the drain on natural resources is likely to be negligible. However, popular species are often grossly over-collected and, as a result, an unreasonable strain has been placed on wild populations. In an attempt to minimise the effect of such exploitation, certain species of tarantula have now been afforded protection under CITES (Convention On International Trade In Endangered Species). Effectively, this means that the trade in these species is now controlled. At the time of printing, all *Brachypelma* species (to which the popular Red-Knee belongs) are CITES listed.

The advantages of purchasing a captive-bred tarantula far outnumber those of buying an imported specimen and captive-breeding should be encouraged wherever possible.

The main advantages are:
1. It is easier to determine the age of a captive-bred specimen.
2. It is likely to be free from parasites, both internal and external.
3. The chances of it having been correctly identified are increased.
4. Often, captive-bred individuals are less aggressive than their wild-caught counterparts.

POPULAR TARANTULA SPECIES

Perhaps the most rewarding aspect of keeping tarantulas is watching a minute, fragile spiderling progress through a series of moults and gradually develop into a resplendent adult. This process may take several years, depending on the species, but the owner's tireless endeavours will ultimately come to fruition.

It is impossible to describe anything other than a fraction of the tarantula species available to the hobbyist within the confines of this guide. I have concentrated on a handful of the most commonly encountered species, including those that are particularly suited to the beginner.

I have used both the common name and the scientific name of each spider described because common names are often of little use and frequently lead to confusion. For example, the Honduran Curly Hair (*Brachypelma albopilosa*) is sometimes referred to as the Wooly Tarantula or Black Velvet, depending on which book you read or dealer you ask. By using scientific names this confusion is minimised.

A male Chilean Rose (*Grammostola spatulata*). The blunt-ended pedipalps and gangly appearance of males easily distinguishes the sexes in adults.

The Chilean Rose (*Grammostola spatulata* and species)

The current favourite 'pet' tarantula, and for good reason because this species is not only attractive and docile but very tolerant of a wide range of environmental conditions. The Chilean Rose is a uniform golden brown/buff colour with a distinct pink iridescence on the cephalothorax.

There is a certain degree of variation

A large and healthy specimen of *Grammostola spatulata*. This long-lived, hardy and docile species is the ideal 'starter' tarantula. Note the spinnerets at the rear of the abdomen exuding silk.

between the appearance of individuals but it would appear that several species of *Grammostola* have been imported under the same name, following incorrect identification. The Chilean Rose (*G. spatulata*) is sometimes incorrectly sold as *Phrixotrichus auratus, roseus* or *cala*.

The Chilean Rose is a slow-growing and long-lived tarantula. Spiderlings take several years to mature, consequently the majority of specimens offered for sale have been wild-caught. The Chilean Rose remains one of the least expensive tarantulas to purchase and is probably the most frequently encountered species in pet shops. There is currently concern over the number of specimens being collected from the wild so the purchase of a captive-bred specimen is to be strongly encouraged. Originating from semi-arid, scrubland areas the Chilean Rose requires a moderately dry (but not desert) set-up. It is a ground dwelling species living in shallow burrows and therefore floor surface area is of greater importance than height when selecting a container. The terrarium should contain a deep substrate and pieces of driftwood or cork bark for cover. An ambient temperature in the mid 70°(F)s is adequate and moderate humidity in the region of 70% is ideal.

Other species sometimes seen include:- (1) the Chilean Beauty (*Grammostola cala*). Although similar in appearance and often confused with *G. spatulata*, this spider has an overall body colouration of a rich fiery red. (2) The Tawny Red (*Grammostola pulchripes*) is another commonly seen species. It is smaller in size than the *spatulata* and dark brown in colour with cream/orange bands around the leg joints. Both *G. cala* and *G. pulchripes* require a more humid environment than *G. spatulata*.

The all-time favourite tarantula - the Mexican Red-Knee (*Brachypelma smithi*). Although expensive to buy, this attractive and hardy species is ideal for the beginner.

The Mexican Red-Knee (*Brachypelma smithi*)

The Mexican Red-Knee is perhaps the most instantly recognisable of all tarantulas. The Red-Knee is a large spider, often attaining a leg span of 6" or more. The carapace is dark brown to jet black in colour, graduating to beige/orange at the edges. The abdomen is a rich velvety black with protruding orange hairs but the most distinguishing feature - and the one that gives this species its name - are the vivid orange/red/yellow leg bands.

With its striking colouration and docile temperament, the Red-Knee is an ideal beginner spider. Unfortunately, the Red-Knee's popularity has resulted in its wholesale exploitation and, consequently, the trade in wild-caught individuals is now severely restricted. Today, specimens offered for sale will almost certainly be of captive-bred origin. They remain one of the more expensive tarantulas to buy because they are not particularly easy to breed, are fairly slow to mature and there is considerable demand for them.

The Red-Knee originates from the arid scrubland areas of Mexico where it lives in deep burrows. The habitat in which it lives is very dry. However, the humidity inside the burrow is considerably greater than that of the outside environment. This should be born in mind when trying to emulate conditions in captivity. Under no account should this species be maintained in a desert enclosure decorated with cacti. While the terrarium should be moderately dry, the spider should always have access to water and a generous substrate in which to burrow.

A temperature in the region of 78°F and humidity of 70-75% is recommended.

If housed correctly, the Red-Knee is a very long-lived species and a 'must' in any serious spider enthusiast's collection.

The Mexican Blonde (*Aphonopelma chalcodes*)

Once commonly seen in pet stores, this species has now become somewhat difficult to obtain and expensive. This is a shame because the Mexican Blonde is an attractive species that is not only fairly docile but also undemanding in captivity.

A juvenile Mexican Blonde (*Aphonopelma chalcodes*) in defence posture.

The Blonde is a medium sized spider with a background colour of pale brown with golden/beige hairs. It is also of a fairly nervous disposition and able to move quickly and shed hairs if distressed.

Originating from the arid regions of Texas and Mexico, the Blonde requires a relatively dry terrarium with a generous substrate for burrowing. A temperature in the region of 75-80°F seems acceptable with good ventilation.

In common with all spiders, it is essential to provide a water dish. A humidity gradient within the enclosure is also beneficial. A truly arid, desert-style terrarium is rarely suitable for any species.

The Curly Hair (*Brachypelma albopilosa*)

The Curly Hair is an attractive, medium sized spider in which the profusion of hair gives the impression of greater size. Body colouration is dark brown to black in newly moulted individuals with longer and distinctly curly protruding beige hairs on the legs and abdomen. The Curly Hair is essentially quite a peaceful species but can be fast moving and somewhat erratic.

Originating from the jungles of Honduras, Costa Rica and Nicaragua, this species requires a warm and relatively humid environment, a deep substrate for burrowing and humidity of about 70-80%.

The Curly Hair (*Brachypelma albopilosa*) is a popular terrarium subject and very suitable for the beginner. A half grown captive-bred juvenile is pictured here.

The Curly Hair is an excellent beginner tarantula. Being inexpensive to purchase, relatively docile, easily bred, straightforward to rear and fast growing, it comes very highly recommended.

The Red Rump Tarantula (*Brachypelma vagans*)

A medium sized, robust species that rapidly reaches maturity. The red rump has distinctive colouration: black in colour with long, vivid red hairs extending from the abdomen, hence its name.

This is not a particularly aggressive species but it could not be described as being docile and has a tendency toward fast and unpredictable movements. Yet this species remains a favourite with newcomers to the hobby and juveniles are easily and inexpensively obtained. Originating from Honduras, Guatemala and neighbouring areas, the Red Rump requires a jungle-like environment and would benefit from a deep substrate in captivity. Care as for *B. albopilosa*.

The Pink Toe (*Avicularia* species)
With many tarantula species, identification can be a problem. The *Avicularias* are no exception with many species sharing almost identical colouration and traits. Fortunately for the enthusiast, most *Avicularias* can be maintained in much the same way.

The *Avicularias* are typically very hairy, fast moving, arboreal spiders. *Avicularia avicularia*, the common Pink Toe, is possibly the most frequently encountered although correct identification is in many cases questionable.

The Pink Toe is a medium sized spider and newly moulted individuals are jet black in colour with distinctive pink feet. There are also red hairs on the sides of the abdomen and a greenish iridescence on the carapace. This is a very manageable and usually quite docile species that is certainly suited to the beginner.

The Martinique Red Tree Spider (*Avicularia versicolor*) is arguably one of the most beautiful of all tarantulas. When seen in daylight, juveniles are a striking turquoise. Adults, however, are a rich plum red colour.

Perhaps one of the most strikingly beautiful *Avicularias* is the Martinique Red Tree Spider (*Avicularia versicolor*). In the juvenile stages of development, and when seen in natural daylight, this spider is a brilliant turquoise with deep blue and grey banding on the abdomen. With maturity, the coloration changes dramatically (hence the name *versicolor*): the blue is lost and adults take on a deep plum red abdomen, reddish-purple legs and metallic green carapace. This is a truly stunning spider.

The Pink Toe and other *Avicularia* species should be housed in tall terraria furnished with a couple of branches to which the spider's tube web can be anchored. Providing that the terrarium is of a reasonable size and furnished with suitable retreats, the Pink Toe can be maintained in small groups. However, it is important that individuals are of a similar size and there is an abundance of food.

The terrarium can be decorated with artificial plants, branches and pieces of cork bark tube. *Avicularia avicularia* enjoys a tropical rain forest environment and therefore the humidity in the terrarium should be in the region of 80% and maintained at a temperature of 78°F. To achieve this level of humidity, and to provide drinking water, the terrarium should be sprayed on a regular basis.

The Salmon Pink (*Lasiodora parahybana*)

This is a seriously impressive spider! Attaining an enormous leg span and being of fairly bulky build this species is, without question, a member of the giant league.

The body colouration is a uniform black/dark brown with prominent pink hairs extending from the legs. There is also a dense patch of pink at the base of the abdomen which, in turn, is sparsely covered with long pink hairs.

The Salmon Pink is a beautiful spider and is easily reared in captivity. Spiderlings can be obtained very inexpensively and grow rapidly, females reaching maturity in approximately four years. The Salmon Pink has a voracious appetite and adult specimens require a diet of appropriately large food items; large cockroaches, locusts and even pink mice will be considered.

Originating from the rain forest regions of Brazil, the Salmon Pink requires a spacious and humid (80%), terrarium suitably furnished with a deep substrate and adequate cover.

While the Salmon Pink is not unduly aggressive in nature, it could not be described as being placid and caution is advised. Otherwise it is a good choice for the beginner looking for a real 'giant'.

The Salmon Pink Bird Eater (*Lasiodora parahybana*) is a truly 'giant' spider and is readily available as captive-bred spiderlings.

The Trinidad Chevron (*Psalmopoeus cambridgei*)

The Trinidad Chevron is another fast moving and somewhat nervous arboreal species. I have included the Chevron because it was the first spider I ever owned.

The Chevron is a beautiful olive green/brown spider of elegant proportions. In natural daylight, vivid orange iridescent markings can be seen on the tibia of the anterior legs and dark stripes on the abdomen. The carapace also has a greenish sheen. The Chevron is typically seen resting with its legs held tightly together, extended to the front and rear of the body whilst lying flat against bark or the terrarium wall. The Chevron is a tube web builder and a healthy specimen will construct a thick silken structure.

Being arboreal in nature, the Chevron requires a tall terrarium furnished with twigs and branches. Humidity should be on the high side (75-80%), with an ambient temperature of 78°F . I know of no other spider that grows quite as quickly if kept well fed; males reach maturity in a staggering 12-18 months.

Although fast moving, the Chevron is not especially aggressive; however, a bite is reportedly quite painful. This species is ideally suited to a keeper that has gained some prior experience of fast moving spiders, but the Chevron remains an undemanding subject in captivity.

The Goliath Bird Eater (*Theraphosa blondi*)

The Goliath is not only the most spectacular of all spiders but, in my opinion, one of the most awe inspiring of land invertebrates in general. One really has to see these monsters in the flesh in order to appreciate their sheer size. With a potential leg span exceeding 10" and an abdomen the size of a hen's egg, *Theraphosa blondi* is the world's largest living spider.

Another interesting trait

The Goliath Bird Eater (*Theraphosa blondi*). This giant is fast moving, somewhat aggressive and able to kick copious quantities of urticating hair.

is the ability this spider has to hiss when disturbed. This process, which is shared with a number of other species, is known as stridulation. Stridulation involves the rubbing together of the mouthparts to make a threatening, hissing sound which is sometimes used in defence.

T. blondi has an appetite to match its size, often consuming large cockroaches and locusts on a daily basis. Pink or fluff mice dangled on a thread in front of the spider are suitable alternative foods.

This species is, however, totally unsuited to the novice. Not only is *blondi* enormous, fast moving, aggressive and able to kick copious quantities of extremely irritating hairs, it is also somewhat sensitive to unfavourable environmental conditions.

T. blondi requires a spacious terrarium with a deep, damp substrate of vermiculite and partially buried cork bark or driftwood for cover. A large water dish is required as this species is prone to dehydration. A temperature of 80°F and humidity of 80-85% is recommended.

Imported specimens are often in poor condition and invariably possess bald abdomens. They can also prove difficult to acclimatise. Thus it is advisable that a captive-bred juvenile is purchased; not only is this more environmentally acceptable but the age and origin of the spider can be determined. However tempting it may be to own a Goliath, it is essential that prior experience of more manageable species is gained before attempting to keep one of these giants.

The South American Bird Eaters (*Pamphobeteus* species)
Numerous species of large theraphosids hail from the tropical rain forest regions of South and Central America. The genus *Pamphobeteus* are reasonably well represented in captivity and frequent imports, notably from Peru, ensure a regular supply. Although somewhat of a generalisation, *Pamphobeteus* are notoriously fast moving, aggressive and, (with a few exceptions), usually of a uniform black/brown colour.

Pamphobeteus require a warm, humid enclosure with ample cover and access to fresh water at all times.

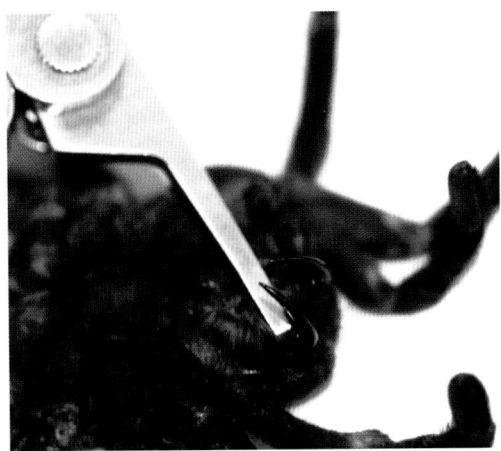

This picture clearly shows the exposed fangs of a South American Bird Eater (*Pamphobeteus* sp.).

Costa Rican Zebra (*Aphonopelma seemanni*)
Originating from Costa Rica and neighbouring regions, the Zebra

is a moderately large spider, charcoal grey to black in colour and covered in longer, lighter coloured hairs. The legs exhibit prominent cream stripes which are particularly eye-catching in freshly moulted individuals. The underside of the spider is a delicate salmon pink colour.

The Zebra is one of my favourite spiders. It is an excellent species for a beginner to start with, being attractive, easy to manage and growing reasonably quickly. Although it has become more difficult to obtain since the export ban in Costa Rica, dedicated enthusiasts and breeders have ensured that this species will not vanish from captive collections. Through established breeding programmes, juveniles are often available and can be purchased fairly inexpensively.

The Zebra requires a rain forest-type set up with a deep substrate for burrowing. A temperature of 78°F and humidity of 80% are recommended.

The Indian Ornamental (*Poecilotheria* species)

The genus *Poecilotheria* embraces some of the most spectacularly marked and beautiful representatives of the family theraphosidae.

The most commonly available species, *Poecilotheria regalis* (the Indian Ornamental), is one of those spiders that every tarantula enthusiast has to own. The Indian Ornamental is a large, arboreal species with stark, contrasting markings of pale grey, fawn and black. The abdomen is dark grey with a broad, pale grey central stripe from which fine dark bands radiate. When provoked, this species will rear up and flash the vivid yellow and black undersides of its front legs.

The Indian Ornamental is a popular terrarium subject and is now being bred in relatively large numbers to satisfy demand. If provided with a regular food supply and a suitable environment, this species matures rapidly and can reach maturity in as little as eighteen months to two years.

Poecilotheria are arboreal in nature and, consequently, their terrarium should be a tall structure furnished with branches, cork tubes and, possibly, artificial plants.

Many *Poecilotheria* possess vivid yellow undersides to their front legs. These are sometimes flashed in defence to startle a would-be predator.

Members of the genus *Poecilotheria* are fast moving arboreal spiders. Although not difficult to maintain, their venom is more toxic than most other commonly kept species and caution is advised.

A temperature in the region of 75-80°F is ample and humidity of 70-80% is preferred.

Other species of *Poecilotheria* have become more widely available of late, including *P. striata*, *P. formosa* and *P. fasciata*. All require similar care.

Along with other *Poecilotheria* species, the Indian Ornamental is extremely fast moving. The venom of *Poecilotheria* is reportedly among the most potent of the commonly kept genera and one would be wise to exercise caution when dealing with these spiders.

The Baboon Spiders

A large number of African Baboon spiders are available to the hobbyist through specialist outlets. Space precludes anything other than a mention but it would be fair to say that many of the Old World 'baboons' are extremely aggressive and a novice would be well advised to steer clear of this group until further experience has been gained.

Among the more commonly encountered species are the Mombassa Golden Star burst (*Pterinochilus murinus*); the very attractive Arboreal Orange (*Pterinochilus spinifer*); the Horned Baboon (*Ceratogyrus darlingi*) with a distinct horn extending from the carapace; and the Giant Rusty Red (*Hysterocrates gigas*), a large, chestnut brown species originating from the equatorial jungles of West Africa. Occasionally one may be offered the Kinani Rusty Red (*Citharischius crawshayi*) or the Hercules Baboon (*Hysterocrates hercules*). The latter are Africa's largest spiders and certainly two of the most beautiful.

All of the species mentioned are aggressive in nature, rearing up and exposing their fangs at the slightest provocation. If the intention is to buy one of these spiders it is important that the correct information, especially regarding habitat type, is obtained in order to simulate conditions in captivity - not all Baboon spiders originate from arid areas!

The Thai Black (*Haplopelma* or *Melopoeus minax/albostriatus* species)

Since the export of live animals was banned from Thailand, the Thai Black has become quite scarce in captivity. Once commonly available and inexpensive, specimens of *minax* and *albostriatus* are now generally captive-bred and much sought after. Recently, another species of *Haplopelma* (possibly *Haplopelma schmidti*) has been arriving from Vietnam in some numbers.

The Thai Black (both *minax* and *albostriatus*) are medium sized spiders with a neat, velvety appearance. The latter has distinct cream banding on the legs and is slightly smaller overall, but otherwise the two species are similar and often mistaken for each other. The abdomen is striped with fine dark lines radiating from a broader, central longitudinal band. These markings, common to many of the *Haplopelma* species, has given rise to their other name - the Earth Tigers (not to be confused with the Central American genus *Cyclosternum*).

This Haplopelma *sp. from Vietnam is a fast moving and aggressive species, a trait shared by many of the Asian* Haplopelmas.

Another species of *Haplopelma* which is sometimes encountered is the Burmese Cobalt Blue (*Haplopelma lividus*). This is a striking spider, similar in appearance to the above species of *Haplopelma* but, when seen in sunlight, the legs of newly moulted individuals assume a shade of deep blue. While it is desirable to own one of these spiders, their highly aggressive nature precludes them from a beginner's collection.

Members of the genus *Haplopelma* are invariably very aggressive in nature, although captive-bred specimens seem to be of a much calmer disposition than their wild-caught counterparts. In captivity they require a humid, jungle-type environment with access to an open water dish at all times. A deep substrate with partially covered cork tubes will facilitate burrowing and provide cover. Sphagnum moss can be used to increase the humidity. A temperature in the region of 75-80°F and high humidity (80%) is required.

HOUSING AND EQUIPMENT

It is not difficult to cater for a tarantula's agoraphobic lifestyle and in captivity most tarantulas require surprisingly little space in order to thrive.

The Terrarium

In their natural habitat, many species dwell in small, dark burrows or will set up home in a tight crevice, enjoying the security of being almost totally enclosed. Therefore the terrarium need not be an elaborate affair; indeed, a tarantula is likely to feel lost and uncomfortable in a large enclosure.

Ultimately the type of terrarium and its proportions are determined by the size and nature of the spider it is to contain. Arboreal species require height whilst ground-dwellers need floorspace. Spiderlings and small juveniles can be adequately housed in clear film pots (or similar containers) if ventilation holes are punched in the sides. They can then be moved to correspondingly larger quarters as they grow.

Pet tanks with ventilated lids are an ideal way of housing tarantulas. Tall terrariums are suitable for arboreal species but should not be used for heavy bodied, terrestrial tarantulas as a fall from one of the walls could result in injury.

The widely available and extremely versatile pet tanks seen in pet stores provide excellent homes for tarantulas. These tanks are lightweight, available in a convenient range of sizes and have a tight fitting, ventilated lid.

Small glass aquariums and specially constructed 'spider tanks' with drop-in glass lids and ventilation panels are equally suitable. Many keepers favour the use of opaque plastic lunch boxes (with breathing holes) for housing their spiders; these are particularly suitable for ground-dwelling species requiring high humidity. Lunch boxes have the added bonus of being very inexpensive and easily stacked. When space is at a premium this method of housing is ideal.

Small glass tanks with drop-in lids make excellent homes for tarantulas.

Most adult tarantulas can be comfortably housed in a terrarium measuring 12"x12"x12" (LWH). Particularly large species will require extra space and arboreal tarantulas would benefit from increased height.

As a general rule, all tarantulas should be maintained individually. There are a few exceptions: *Avicularias* and some *Poecilotheria* can be housed in species specific groups if provided with ample space for retreat. However, to avoid cannibalism, it is probably safest to house them separately.

Terrarium Furnishings

The terrarium should be furnished with a generous ground covering. Vermiculite is almost universally acceptable and is suitable for use regardless of whether an arid or humid environment is required.

Vermiculite is easily obtained from specialist suppliers or from garden centres and building merchants. It is an inert mineral substance that has a massive water holding capacity. It is also very light and easily moved around by the spider when in pursuit of burrowing activities. Peat, coir and bark chips are also suitable substrates but, for practicality, it is hard to beat vermiculite. Sand and gravel are of limited use.

The depth of the substrate is of lesser importance for non-burrowing species but for spiders with a predilection for earth-moving activities, a layer of 3-6" is recommended.

Other terrarium furnishings can be used to provide additional cover; cork bark tubes, driftwood and artificial plants can be used as retreats and for aesthetic reasons. Avoid using heavy or sharp pointed objects such as large pieces of wood or rocks because these may be dislodged and cause injury to the spider. Real plants, particularly cacti, should not be considered. Aside from being unlikely to survive under the low lighting levels inside the terrarium, they will probably be uprooted or disturbed by the spider. It is common to see photographs of cacti being used to

Tarantulas should always have access to water.

decorate the terrarium interior. Due to the prickly nature of these plants it is possible for the spider to injure itself on the sharp spines.

Aside from a suitable substrate and some form of cover, the only other essential additions to the terrarium are a water dish, a humidity meter (hygrometer), thermometer and possibly some form of heating. All tarantulas must have access to water. A small, shallow water dish (petri dishes or jam jar lids are ideal) should always be made available. In the case of juvenile spiders being maintained in the likes of film pots, and where space precludes the provision of a dish, a small piece of wet cotton wool can be used. This will need to be re-hydrated/replaced regularly.

Many arboreal species prefer to drink water droplets from the surface of their web and therefore frequent spraying is an essential requirement.

Heating and Lighting

As far as heating is concerned, most tropical spiders will require some supplementary warmth over and above that of normal room temperature. This is best achieved with the use of a low wattage heating mat or cable. These can be attached to the rear or side wall of the terrarium - either internally or, better still, on the outside. If the heater is placed underneath the tank, as is sometimes suggested, only part of the ground should be heated. However, the spider's natural instinct to burrow may be disrupted as it would effectively be burrowing towards the heat source rather than escaping from it. There is also the danger of the mat overheating because a deep layer of vermiculite is an excellent insulator.

The advantage of such heaters is that multiple tanks can be inexpensively heated by one piece of equipment. A thermostat can be connected to regulate the temperature more accurately. Whether a thermostat is used or not, the desired temperature should be attained *before* introducing the occupant.

The use of incandescent light

When seen in daylight, many tarantulas assume a colourful iridescence; often these colours are lost under artificial light.

bulbs as a method of heating is largely unsatisfactory. Apart from the excessive light output, there is the danger that the spider may burn itself on an unprotected bulb. If this method of heating is to be considered, then a blue or red bulb should be utilised as these colours cannot be detected by the tarantula's poor eyesight.

As a rough guide, most tarantulas appear to thrive in a temperature range of 75-80°F with a slight temperature drop at night. It is beneficial to create a temperature gradient within the terrarium, although when dealing with such small enclosures this is rarely practical.

Tarantulas are nocturnal and will go to great lengths to avoid bright light. Therefore, in captivity, the tarantula should be provided with indirect, subdued illumination and terrariums need not be individually lit. Always remember that a glow, as opposed to a glare, is preferable. Positioning the terrarium on a windowsill is a recipe for disaster. Not only is it likely to be too bright but there is the danger of the tank dangerously overheating.

Humidity and Ventilation

A spider's well-being is entirely dependent on the keeper providing the right environment. A keeper therefore has a moral obligation to ensure that any spider under his or her care is properly looked after.

Perhaps one really only becomes aware of just how important it is to maintain the right environmental conditions when a spider suffers complications whilst moulting, at which time the level of humidity is crucial.

I have noted the optimum humidity for each species of tarantula described:- those originating from rain forest environment generally require humidity in the region of 80% and those from arid areas require about 50-60%. It is important to question the dealer or breeder on this point before a prospective purchase.

Damp vermiculite and sphagnum moss in combination with regular spraying can be used to achieve high humidity. Shallow water containers filled with the above are also useful. It may also be necessary to mask some of the grills on commercially produced pet tanks with tape to prevent the terrarium from constantly drying out. When striving to create a humid environment it is important not to overlook ventilation; a stagnant terrarium should be avoided at all costs. If a lunch box type container is being used, don't forget to punch numerous holes in the sides.

Day to day maintenance

Tarantulas produce very little in the form of solid waste. Consequently, their terrarium rarely needs to be cleaned out. However, uneaten or partially eaten food items should be removed on a regular basis. A poorly maintained terrarium will become a breeding ground for mites.

Aside from regular checks on the spider's well-being, routine maintenance should include: spraying, replacement of drinking water, feeding, temperature and humidity checks.

FEEDING

Tarantulas require live food, or at least they must be able to detect the movement of any food items they are offered. The diet of most captive spiders is likely to comprise locusts, crickets, mealworms and cockroaches. The size of the food items offered should be directly proportional to the size of the spider. Large specimens will have no difficulty in tackling an adult locust or pink (new born) mouse, if dangled on a thread.

Locusts, house crickets and cockroaches are all excellent foods. However, special care should be taken when offering large locusts, black field crickets or any oversized food insects because of the danger that the intended food item may attack the spider. Injury is most likely to happen when the spider is defenceless during, and just after, a moult.

Live foods are easily obtained from pet shops that sell reptiles.

It is beneficial to offer a variety of different foods. Vitamin and mineral supplements are on the whole considered unnecessary; if the spider is fed on insects which themselves have been provided with a rich and varied diet, their nutritional content will be far greater. Mealworms can be used as a supplementary foodstuff and are particularly useful as a standby when other foods have been exhausted. However, due to their poor nutritional value, they should not be used as a staple food.

Small juvenile tarantulas should be fed two or three times per week. Large juveniles and adults need only be fed weekly, offering one or two food items at a time. Newly emerged spiderlings will not feed initially, only after they have moulted. Acceptable first foods include micro-crickets and fruitflies (*Drosophila*), all of which can be purchased by mail order from specialist outlets.

The appetites of different species and amongst individuals can vary enormously but there is no danger of overfeeding your spider. The more it eats, the faster it will grow and the eventual size of the adult will, in part, be determined by its diet.

Tarantulas are, in the main, fairly inactive creatures and consequently energy expenditure is minimal. There are reports of tarantulas going without food for periods as long as two years. While this is not to be recommended, a healthy spider will not come to any harm if left unfed for several weeks. When approaching a moult, it is common for a tarantula to enter a sometimes lengthy period of fasting, so do not be unduly concerned if your spider refuses to feed. A low temperature will cause the spider to be sluggish and lose interest in food. Intense lighting may also have a bearing on appetite. If a spider refuses to feed, check the environmental conditions and try offering an alternative food item. Above all, don't panic!

MOULTING AND SEXING

Spiders, in common with all arachnids and other invertebrates possessing a ridged exoskeleton, can only grow when their tough outer layer (or cuticle) is soft and flexible.

Apart from variations in the size of the abdomen, a spider needs to moult in order to grow, discarding its old, hardened cuticle and moulting into a fresh, new, flexible one.

This newly moulted Peruvian 'bird eater' exhibits striking colouration. Prior to the moult, it was a dull black and brown.

Ecdysis (moulting) is the most risky period in a tarantula's life. It is immobile, defenceless and therefore vulnerable to attack. In addition, if the environment is not right, the spider may be unable to discard its old skin, and become trapped. These photos illustrate the process of ecdysis, ending with the free spider lying on its back, and the cast-off skin.

MOULTING AND SEXING · YOUR FIRST TARANTULA

The discarded skin of *Avicularia urticans*. Note the way in which the cephalothorax has split around the edge and the carapace has lifted off, allowing the spider to pull itself free.

This process, known as ecdysis, is fascinating to observe. Just prior to moulting the spider will often spin a soft, flat web on which to rest. It will lie perfectly still on its back in the centre of this web. The old skin will split around the edge of the cephalothorax and abdomen, the carapace will lift off like a lid and the spider will slowly ease itself free.

The emergent spider will be practically immobile, its legs will be soft and bendy and its fangs dysfunctional. The first thing one notices about a newly emerged specimen is the white fangs, white leg joints and intensified colouration. If the spider originally possessed a bald abdominal patch, any lost hairs will now have been replaced. The period during and just after a moult is the most critical in a tarantula's life. The spider will be defenceless and vulnerable to attack.

As mentioned earlier, a successful moult is heavily dependant on environmental factors. If humidity is too great or conditions are too dry, the spider will become trapped in its old skin or emerge as a twisted cripple. It is very distressing for the keeper (not to mention the spider) when a moult goes wrong and the spider becomes severely trapped in its old skin.

Should a spider become trapped, a certain degree of success can be attained by applying a glycerine solution with a paintbrush to any caught limbs. This method was first described in Al David's book *The Complete Introduction to Tarantulas 1987*. If the spider is severely caught then there is little that can be done.

It is ironic that the hardiness and adaptability of some tarantula species can ultimately result in their demise. By seemingly thriving in substandard conditions, the keeper can be lulled into a false sense of security. Often environmental problems only become apparent when disaster strikes during a moult and, invariably, it is then too late to rectify the situation.

There are a number of warning signs displayed by the tarantula just prior to moulting:
1. The spider may have refused food for some time.
2. The abdomen will have become enlarged, swollen looking and darker in appearance.
3. A thick, flat web may have been constructed.

Adult and sub-adult tarantulas moult fairly infrequently, maybe every six or twelve months. The time lapse between moults is heavily dependent on a number of factors (such as species, diet and environmental conditions). Juveniles, on the other hand, moult much more regularly - sometimes every other month or so.

It is important not to disturb your spider just prior to, and just after, moulting. It may take a day or two for the new skin to harden and at this time the spider will be extremely fragile and easily damaged. Food insects should be removed from the terrarium if a moult is imminent.

Sex Determination

Many people claim to be able to distinguish between male and female juvenile tarantulas. Without a shed skin to examine it is virtually impossible to tell the difference; even then, it takes an expert's eye, a shed skin and a microscope. Adults, however, are easily distinguished. Males possess swollen and blunt-ended pedipalps and often the tibia of the first set of legs exhibits a distinct spur or hook, although this characteristic is not shared by all species. The overall proportions of the spider also differ, the male of the species being of more slender build and often having a somewhat gangly appearance.

There are frequently dissimilarities in the colouration of the male and female, colours often being more intense in the male. In some species, the Malay Feather Leg (*Coremiocnemis validus*) for example, the sexual dimorphism is so great that the sexes could easily be mistaken for being different species.

Record keeping

It is good practice to keep record cards for each spider. This will be of particular use if breeding is going to be attempted at a later date. By recording the date the spider was purchased or hatched, the frequency of feeding and making a note of each moult, a detailed profile of the spider can be constructed.

HEALTH MATTERS

On the whole it is relatively straightforward to distinguish between a healthy tarantula and one that is in poor condition. The abdomen should be plump, well rounded and free from any irregular bulges. Bald patches on the abdomen indicate that the spider has been kicking urticating hairs possibly, although not always, as a result of rough handling. The legs should comfortably support the spider's body and bowed or twisted legs indicate a previously problematic moult.

If the spider's legs are curled underneath the cephalothorax, or if it moves in an arthritic fashion, it is likely that the spider is very dehydrated or is suffering from a severe ailment.

A tarantula that is approaching moult will often have a dull, tatty appearance. Providing that it appears otherwise healthy, there is no reason why it wouldn't make a perfectly good purchase.

A rather tatty looking adult female Goliath Bird Eater (*Theraphosa blondi*).

Missing and damaged limbs

It is quite common to see tarantulas with one or more legs missing. This is particularly true of wild-caught spiders, which sometimes arrive in poor condition.

Most tarantulas cope with the problem of a missing limb remarkably well. If a leg is damaged, the spider will often remove the whole leg from the point at which it is attached to the cephalothorax. Since a tarantula's blood or haemolymph contains no coagulating agents, it will continue to bleed unless the limb is removed at this joint which acts as a natural cut-off point.

A missing leg will be regenerated at the next moult. The replacement leg will be conspicuously smaller than the others at first but over successive moults any difference will become less obvious.

Haemolymph loss can be minimised with the administration of plain talcum powder or icing sugar to the damaged area (Web 1992).

Mites and other parasites

The most frequently seen parasites on tarantulas are mites. These creatures often congregate around the edges of the carapace and between the leg joints. This is usually only a problem with wild-caught specimens. Imported Baboon Spiders seem particularly prone to infestations.

Obviously one cannot use insecticides: even those designed for use with reptiles should be avoided. The only way of ridding the spider of these parasites is to painstakingly remove them with tweezers or attempt to lift them from the spider with a fine paintbrush charged with petroleum jelly. Clearly this is impractical when dealing with an aggressive tarantula and one may have to wait until the spider moults, hoping that the mites will be discarded with the old skin. In any case, a thorough cleaning of the terrarium and replacement of the decor is advised.

Other parasites sometimes encountered are parasitic wasps and flies which lay eggs inside the tarantula's abdomen. The eggs hatch and the emergent larvae will feed on the living spider. There is currently no cure for this ailment and if evidence of parasitic wasps becomes apparent the spider should be destroyed. This affliction is usually confined to imported specimens.

Dehydrated Individuals
Newly purchased or imported tarantulas may be dehydrated on arrival. If a spider becomes severely dehydrated it is advisable to place a large shallow dish filled with saturated cotton wool in the terrarium. Position the spider in the dish and cover the tank with a cloth to minimise disturbance.

Above and below: by far the cheapest, most rewarding and environmentally friendly way of building a tarantula collection is to purchase captive bred spiderlings or juveniles from a reputable breeder or dealer.

Longevity
It is difficult to give accurate information concerning the life span of different tarantulas. However, it is known that many species, if provided with the right environment, may live in excess of twenty years in captivity. It is also fair to say that slow growing species can be expected to outlive those that mature rapidly and that females live considerably longer than their male counterparts.

When selecting a spider as a pet, males are best avoided for this reason. Large juveniles or sub-adults are preferable to particularly large, dull coloured specimens which may be nearing the end of their life.

FINAL REMARKS

Defensive posture. Just because a species is broadly described as being docile, it doesn't mean that certain individuals are not bad tempered!

Tarantula keeping is still in its infancy. Because of this, there is still much to learn concerning the best methods of maintaining and breeding different species. This book has attempted to provide the beginner with a sound introduction to what can be a fascinating and extremely rewarding hobby, but there will always be room for the refinement of husbandry methods and both common and scientific names are subject to change. As an enthusiast, there is an opportunity to develop and perfect these methods and make genuine discoveries based on the observation of captive specimens.

It should once again be stressed that tarantulas are not domestic pets. They are, and will remain, wild animals capable of causing injury. It should also be remembered that some people are genuinely terrified of spiders and owners of these creatures must therefore ensure that a responsible attitude is adopted at all times.

BIBLIOGRAPHY

Baxter, RN 1993
Keeping and Breeding Tarantulas
Chudleigh Publishing

Browning, JG 1989
Tarantulas
TFH Publications

Coote, J 1993
Tarantulas.
Practical Python Publications

Rankin, W and Walls, JG 1994
Tarantulas and Scorpions, Their Care in Captivity
TFH Publications

Hancock J and K 1992
Tarantulas: Keeping and Breeding Arachnids in Captivity
R and A Publishing

Preston-Mafham R and K 1993
Spiders of the World
Blandford

Stone, JLS 1992
Keeping and Breeding Butterflies and other Exotica
Blandford

Web, A 1992
The Proper Care of Tarantulas
TFH Publications

Hillyard, P 1994
The Book of the Spider, From Arachnophobia to the Love of Spiders
Hutchinson

Marshall, SD 1996
Tarantulas and Other Arachnids
Barrons

Schultz, SA and Schultz, MJ 1998
The Tarantula Keeper's Guide
Barrons

Vosjoli P de. 1991
Arachnomania
Advanced Vivarium Systems

David, Al 1987
The Complete Introduction to Tarantulas
TFH Publications

Reger, Barbara 1995
Tarantulas as a New Pet
TFH Publications

USEFUL ADDRESSES

Cicada Biological Supply
White Leaved Oak
Bromsberrow
Ledbury
Herefordshire
HR8 1SE
Tel. 01531 650059
For livestock, equipment and books. Advice freely given.

Custom Aquaria
Units D-F
Mark Grove House
Allen Road
Rushden
Northamptonshire
NN10 ODU
For the manufacture of specialist aquaria and terraria.

The British Tarantula Society
81 Philimore Place
Radlett
Hertfordshire
WD7 8NJ

British Arachnological Society
71 Havant Road
Walthamstow
London
E17 3JE

Pet Reptile Magazine
Freestyle Publications
Alexander House
Ling Road
Tower Park
Poole
Dorset BH12 4NZ
A monthly magazine covering many invertebrate and herpetological related topics.

Reptile & Amphibian Hobbyist
TFH Publications
Nest Business Park
Martin Road,
Havant
Hampshire PO9 5TL

The American Tarantula Society
New Mexico State University
South Eastern Agricultural Exp. Station
67 East Dinkus Road
Artesia
NM 88210
America

American Arachnological Society
American Museum of Natural History
Central Park West at 19th Street
New York
NY 10024
America